Lionel S. Beale

The Mystery of Life

an essay in reply to Dr. Gull's attack on the theory of vitality in his

Harveian oration for 1870

Lionel S. Beale

The Mystery of Life
an essay in reply to Dr. Gull's attack on the theory of vitality in his Harveian oration for 1870

ISBN/EAN: 9783337095550

Printed in Europe, USA, Canada, Australia, Japan

Cover: Foto ©berggeist007 / pixelio.de

More available books at **www.hansebooks.com**

THE
MYSTERY OF LIFE.

THE

MYSTERY OF LIFE:

AN ESSAY

IN REPLY TO

DR. GULL'S ATTACK ON THE THEORY OF VITALITY

IN HIS

HARVEIAN ORATION FOR 1870.

BY

LIONEL S. BEALE, M.B., F.R.S.,

*Fellow of the Royal College of Physicians; Physician to King's College Hospital,
and formerly Professor of Physiology and of General and Morbid
Anatomy in King's College, London.*

WITH TWO COLOURED PLATES.

LONDON:
J. & A. CHURCHILL, NEW BURLINGTON STREET.
PHILADELPHIA, LINDSAY AND BLAKISTON.

1871.

PREFACE.

THE groundwork of the following essay was published in the "Fortnightly Review," for September 1st, 1870, in reply to some remarks by Dr. Gull, in his Harveian Oration delivered before the President and Fellows of the Royal College of Physicians, on June 24th of the same year.

The importance of the issue cannot easily be exaggerated. Forced by the evidence of very many facts to accept the theory of vitality, I would, nevertheless, abandon this idea assailed by Dr. Gull, if the truth of any one of the physical doctrines of life opposed to my views had been rendered probable by scientific evidence. So far, however, facts and observations on things living support the idea of vitality, and are not favourable to any mechanical or chemical hypothesis of life yet proposed, which last, therefore, rest at present upon authority alone.

In this essay I have ventured to state some of the arguments, which seem, to my judgment, fatal to all physical hypotheses of life, and have adduced a few of the most important facts and observations, which have led me to advocate a very different conclusion.

GROSVENOR STREET;
    February, 1871.

HARVEY.

"He was used to say, that he never dissected the body of any animal without discovering something which he had not expected or conceived of, and in which he recognized the hand of an all-wise Creator. To this particular agency, and not to the operation of general laws, he ascribed all the phenomena of nature."—"*The Roll of the Royal College of Physicians,*" by Dr. Munk.

THE
MYSTERY OF LIFE.

———o———

THE Harveian Orator of 1870 pays me the compliment of bringing under the notice of the President and Fellows of the College of Physicians some views of mine concerning the nature of life. Dr. Gull differs from me, however, and considers it "strange" that any one should entertain the opinion which I believe to be correct. Would that he had seen the things that have convinced me, or carefully submitted to examination the data upon which my conclusions are founded, and had subjected to critical analysis the facts and arguments I have advanced in favour of the views I feel it necessary to uphold. Had Dr. Gull, and others who differ from me, acted in this way, fallacies might have been detected, errors pointed out, and a more correct interpretation of facts afforded, than I have been able to give. Instead, however, of entering

B

into an examination of the grounds of my theory, Dr. Gull simply accepts, supports, and advocates the views of those who hold that "life" is a form or mode of ordinary force, and attacks the position that life is a power distinct and apart from the forces of the non-living world; but he does not reply to the objections which have been advanced against the theory he so warmly advocates, nor does he overthrow the arguments adduced in support of the doctrine he desires to controvert.

My view, which has been assailed by Dr. Gull, is this: that "life is a power entirely different from and in no way correlated with matter and its ordinary forces." The words have been taken from the second sentence of my work on Protoplasm, which, however, runs thus in the original: " Life is a power, *force, or property of a special and peculiar kind, temporarily influencing matter and its ordinary forces, but entirely different from, and in no way correlated with, any of these.*"

Strange as it may appear in these days, the orator commences his oration by implying that there exists in some minds a doubt " whether man is altogether an object of scientific study or not; whether the mysteries of his organisation are fairly subjects admitting of investigation;" and, therefore, whether it is becoming in the Harveian Orator to stir up our minds to search these mysteries out to their fullest extent. It does not appear that anyone has actually expressed himself against such inquiry, but to Dr. Gull himself are we indebted for the inference that one who entertains the opinion which he desires to controvert must therefore hold life to be no proper object of investigation, and must assume that the phenomena of living beings are "out of the range of science." Such a person, moreover, deplores the orator, consigns us to a "perpetual mental inactivity and ignorance in that region of knowledge in which, above all others, man is interested." But is such an inference natural or just? Does it really flow from the premises, or has it anything whatever to do with them? Might it not have been drawn from any other supposable premises with almost equal justification? For how can the opinion that life is a *power entirely different from ordinary force*, involve the position that man's organisation is not fitted for scientific study? If, Dr. Gull seems to argue,

life *is* correlated motion, it *is* a legitimate subject for scientific inquiry; while, if life is *not*
correlated motion, the changes in the matter
cannot be investigated.

Less than thirty years ago, says the Harveian
Orator, it was gravely questioned (but by whom
is not stated) whether a living body could not
generate some of the elements of which it was
composed by its own vital force, and it was
considered that an organism *formed the materials* of its higher structures, and was capable
of transmuting elements.   Is there not here
just the suspicion of a suggestion intended that
persons who in these days consider life to be
different from, and in no way correlated with,
the ordinary forces of non-living matter, really
may have some strong affection for those
equally (?) absurd views of thirty years ago ?

Surely it ought not to be necessary for me to
state that I do believe the correlation of forces,
as well as the truths of physics and chemistry,
as firmly as any man can believe them.   I
never supposed, not even when I was a first
year's student, that an organism *formed* the
materials of which it was composed out of
nothing or out of itself, or that one element

could be converted into another element. Dr. Gull, like many more who have committed themselves to force views, does not appear to see that my argument is not in any way opposed to facts or to established truths of physics.

Such suggestions as those above referred to seem to me objectionable, if not unjust. They are calculated to make people think that the conclusions I have arrived at are absurd and unreasonable. They may even prejudice people very unfairly against my views. If, indeed, this is not the object of the statements, the reason for introducing such remarks is by no means clear. To attempt to excite prejudice in a reader's mind against the conclusions of an opponent, instead of attacking his facts and arguments, is almost an admission of weakness. It is indeed significant, if, as seems to be the case at this time in England, an investigator cannot be permitted to remark that facts which he has demonstrated, and phenomena which he has observed, render it impossible for him to assent at present to the dogma that *life is a mode of ordinary force*, without being held up, by some who entertain opinions at variance with his own, as a person who desires to stop or retard investigation, who disbelieves in the correlation of the physical forces, and in the established truths of physics.*

Is it possible that belief in a something, a power, force, agency, or, call it what you will, which is beyond the range of physical and chemical investigation, and cannot be rendered evident to the senses, should disqualify a man for scientific investigation, any more than a belief in a God renders it impossible for him to successfully pursue observation and experiment? It ought not to be necessary to state that the proposition that *vital power* is distinct from *force* does not involve a belief in the absurdity that life creates matter or transmutes one element into another.

Whatever may be the fate of the inferences I have drawn concerning the nature of *vital*

* Dr. Tyndall goes even still further. Instead of answering arguments, he gives expression to some of the words of his friend, Huxley, and speaks of me as a "microscopist, ignorant alike of Philosophy and Biology," and as having been "lately Professor in a London College, famous for its orthodoxy." That I am not a convert to the Philosophy and Biology of Tyndall and Huxley is perfectly true ; but that my connection with King's College has in any way influenced my views, is a suggestion as devoid of foundation as the fiery cloud hypothesis of evolution itself.

actions, they have been deduced from facts of observation. The theory has, as it were, forced itself upon me in the course of my work. In the spring of 1861 I had the honour of delivering, at the College of Physicians, a course of lectures " On the Structure and Growth of the Simple Tissues of the Body ; " during the delivery of which, upwards of sixty microscopical specimens were exhibited and described. The conclusions I drew were based upon the facts thus publicly demonstrated. My lectures, with a description of the specimens, were afterwards published and illustrated with numerous drawings from the preparations. This volume was afterwards translated into German by my friend, Prof. Victor Carus, of Leipzig. Most of the original preparations still remain in good preservation, and many new ones have been added, year after year. So far from my conclusions having been weakened by the more recent researches of other observers, they have been confirmed and extended.

The evidence in favour of *vitality* being an agency distinct from mere *force*,—being the power by which all living things are characterized, and which absolutely separates them from the non-living, is so strong that it seems to me we can only escape from the conclusion if we deny or ignore incontrovertible facts.

The particular view adopted by me has then resulted from facts of observation. It is a conclusion which has forced itself upon my mind after many years of careful work,—a conclusion from which I have tried to escape, but have failed to do so. I have endeavoured to account for the phenomena by other theories, but have not been successful, nor have attempts on the part of others been more fortunate. The doctrine of vitality is one which I should never have accepted if, by the views more generally entertained and taught, a sufficient explanation of the simplest phenomena of living beings had been afforded ; if, for example, the movements of the simplest forms of living matter could have been accounted for, if the changes which occurred during the development of a cilium, during its period of vibratile activity, and when it died, could have been explained—nay, if the mode of increase of a blade of grass, or the sprouting of a microscopic fungus had been shown to depend upon physical and chemical changes only.

Notwithstanding all that has been asserted to the contrary, not one vital action has yet been accounted for by physics and chemistry. The assertion that life is correlated force rests upon assertion alone, and we are just as far from an explanation of vital phenomena by force hypotheses as we were before the dis· covery of the doctrine of the correlation of the physical forces. In short, this most important discovery in physics does not affect the question of the nature of the phenomena peculiar to living beings.

Each additional year's labour only serves to confirm me more strongly than before, in the opinion that the physical doctrine of life cannot be sustained, and when I review in my mind the evidence upon which the doctrine of *vitality* rests, it appears to me extraordinary that any one can persuade himself that a thing, possessing in itself the power or property of transforming matter and force in a definite way, is itself mere matter and force,—that that which *converts* is no more than that which *is converted*. Because, as Dr. Gull remarks, "a mechanical cause in its simplest form is evolved into its effect by suppression," it does not surely follow that the organic processes are correlatives of the lower forces.   Because heat, light, electricity, &c., are but other forms or modes of motion, that, therefore, LIFE must be also a higher correlate, is very strange reasoning.   Such an assertion begs the whole question of the analogy existing between the living and non-living.   Because the *living* body does not transmute its material elements, therefore, all vital phenomena result from physics and chemistry, is a very remarkable argument, but surely not conclusive.   We all agree that the *materials* of which the living body is constituted are the same as those of which non-living matter consists, and that the forces are the same.   The question is whether the arrangements of the matter, the form of the living being, and the guidance of the forces in living things, are due to the working of ordinary material forces, or to a power of an altogether different order.

The relation between *vital power* and the ordinary forces of matter may not be more intimate than the relation between the man who makes a water-mill, and the forces which raise the water that drives the wheel, or the materials of which the mill is constructed. And yet the water-mill could not have been made by the water nor by the wood nor iron which in part constitute the mill, nor by the mighty forces imprisoned in these materials. The man, not the forces of the matter or of the water, *constructs* the mill. Where, then, is the evidence that justifies Dr. Gull, and those whom he follows, in asserting that any *form or mode of ordinary force has constructive power ?* Force is mighty, force is powerful, and force may be *destructive;* but what evidence can be adduced in favour of the *constructive* agency of any mode of force ? Can any or all the forms of force yet discovered *construct* an insignificant monad any more than they can make an umbrella or build a house ? Dr. Gull neither notices the objections which have been raised to the view concerning the *forming, building,* and *constructing* powers of force, nor adduces one new fact or argument in its support.

Some force devotees may perhaps be inclined to regard the most beautiful works of art, as well as of nature, as mere force productions, and hold that *form* is but the image impressed by force. But unless something directs, will form appear ? Is not that something other than the force, is it not master and director of the force ? Are not *force* and *matter* his *tools*, and does not *form* result from the particular way in which the *master, director*, and *designer works ?* Whatever name be given to this something, I cannot conceive that it can be a correlate of material *force*. Will any one maintain that the man who made a machine is a correlate of the heat that set it in motion after it was made ?

The term *vital power* has been applied by me to the marvellous agency which, besides giving rise to form, silently effects the analysis of compounds and causes their elements to be rearranged, so that when synthesis occurs new compounds result, which did not exist before. The complex operations of analysis and synthesis are performed as in a moment, and without any bulky, cumbrous, though elaborate and beautiful appliances, such as the chemist and physicist are forced to employ, and the skill to use which can only be acquired after years of patient study and earnest work. Nature's "apparatus" is a tiny mass of clear, transparent, structureless stuff, it may be less than the $\frac{1}{100.000}$th of any inch in diameter. This is also

Nature's laboratory. Here her chemist, "*life*," is at work, and his work is perfect.

But let us consider the matter from another stand-point. Here are two minute masses of perfectly *structureless*, *colourless*, living matter. No difference between them can be demonstrated by physics or chemistry. They have no structure. They are soft and diffluent. One placed under certain conditions will become a dog, the other a man ; but from the dog-germ you cannot by any alteration of conditions obtain a man, any more than from the man-germ anything but a man, or parts of a man, can be evolved. Now what is the difference between the man-living-matter and the dog-living-matter which could not be distinguished by physical or chemical investigation ? I would answer a transcendent difference,—but in *power*. Dr. Gull would say these germs " became through a definite set of *physical* (!) *relations* like the parents from which they sprang." He remarks, that whether the germs are as "limited and specific as we have hitherto regarded them is the *questio vexata* of the day." But, as will be observed, the whole question is begged in the words "physical relations." The relations in question cannot be correctly called *physical*, for they are very complex, being partly physical and partly vital, while they result in part from the state of things brought about by the action and reaction of vital and physical agencies. The relations in question could never have been established by the operation of physics only.

But the conclusion accepted is that "life" is an *undiscovered correlative* of force. And the *undiscovered*—that is, a mere guess or fancy—is a modern idol. Has science, with her *observation*, her *experimental* method, and her *facts*, really been brought to this?

Can the Harveian Orator adduce good reasons for his "full and implicit belief that the as yet mysterious phenomena of life are correlative with the lower forces of nature?" As soon as this has been done, many who dissent, and can express clearly the grounds of their dissent, will cordially embrace the new faith. But surely we who differ may be pardoned for being heartily tired of hearing the argument repeated, "that because nothing passes into us but *matter*, and nothing passes out of us but *matter*, and nothing can be got from us after we are dead but *matter*—therefore we who are actually

living are *matter* only." Everybody knows
about the matter, but he wants to know what
makes this matter do things while it is alive
which it cannot do when it is dead, and which
matter cannot be made to do before it begins
to live,—before it becomes a part of matter
which lives already.

According to physical force doctrines the
living state is not very widely separated from
the non-living condition, and Dr. Gull confesses
that he cannot draw the line between the living
and the dead. But surely it will not be maintained that a line *cannot be drawn*. If, then, one
form of philosophy is confessedly incompetent to
explain a familiar phenomenon, is it not better
to try some other kind of philosophy than to
accept, and without enquiry, the conclusion
that states so very far removed from one another,
so absolutely distinct according to all ordinary
evidence and means of judging, really differ from
one another only in degree ?

It is admitted that the supposed vital correlate has not yet been obtained from heat, light,
electricity, or converted into any one of these
or other modes of ordinary energy, but it is
said this will be proved to be possible. The

force-correlation doctrine of vitality, therefore, draws largely not only upon our faith in the prophecy that some new mode of force *will be* discovered, but that when it shall have been discovered it will undoubtedly transcend in its capacity every mode of force yet known. It will be found to be at least as far above chemical force as this is superior to ordinary motion. But the advocates of the force-doctrine of life stake everything upon the truth of the prophecy and we who irreconcileably differ in opinion from them must be content to wait, and in the meantime be restricted to observation and experiment, because we are not gifted with the prophetic spirit, while they, more fortunately circumstanced, may develope fancies, and expound the discoveries of their imagination. But for how long are we expected to continue to bow down to dogma—for a limited period only, or for as long a time, as the prophecy may remain unfulfilled ?

It *may be true* that chemistry ceases in our living tissues under that form to "appear under some higher correlative," but it has not been proved, nor has any step been made towards proof. . Can anybody give us a conception of the "higher correlative" whose coming is announced, or is it a fiction of the mechanical imagination —an adumbration of a phantom in that constantly recurring dream about unity ? Because we have experience that heat may assume the form or mode of motion, light, magnetism, or electricity, that *therefore* there are other modes of force unknown, but which will certainly be discovered some day, and that one of these is the *vital mode*, is the reasoning we are expected to accept. Correlation is the *abracadabra* of· the long prophecied, but still non-existent science of mechanical biology. That the *non-forming* correlatives of non-forming primary energy, heat, light, electricity, magnetism, chemical action *may* have a yet undiscovered correlative *endowed with formative power* cannot be denied, but is it more probable than would be the assertion that intelligence has been evolved from stones, or that order has resulted from chaos, or that design has been the necessary consequence of conditions powerless to condition ?

Now let us place but a portion of one of the lowest living forms under a high. magnifying power, and let us see if the force-correlation·

C

doctrine will enable us to account for the phenomena it exhibits. We demonstrate some transparent stuff which takes from around it certain matters dissolved in the water in which it lives, and converts these into stuff like itself. How ? " By its molecular machinery, worked by its molecular forces," answers the force-philosophy. It moves, and different portions of the little piece move in opposite directions at the very same instant. What makes it move ? " Its molecular machinery by the laws of molecular physics," we are told ! And is it really to be expected that inquiry is to be stifled, and curiosity satisfied by such announcements as these ? How are the forces conditioned ? What is the structure of the supposed force-conditioning machinery, and how did it make itself ? No answer but " future investigation will decide." But at this time we demonstrate, by our own observation, that the stuff that moves is clear, transparent, and, under a power of five thousand diameters, perfectly structure-less. We can see no " machinery," and we know that there is no machinery in the living matter at all like any machinery known to us, or in any way tending to approach it in character. If, therefore, the term "machinery" is to be applied to this transparent matter, the word has had a new meaning assigned to it, and syrup or water might be spoken of as "machinery," and would come into the same category as the so-called "molecular machinery" of living matter. This clear, transparent, structureless living stuff came from stuff like itself, which had similar powers and properties. How this can properly be regarded as the child of conditions, the creature of external circumstances, the offspring of physical force, an outcome of the non-living, it is indeed difficult to understand. Was it not derived from parental living matter? or have our eyes and understandings utterly deceived us? Are its phenomena of motion, of increase, and of multiplication due to the conversion of the forces of the stuff it lives upon, and not in any way to peculiar power or influence transmitted to it from its predecessors, and manifested by them, but by no form of non-living matter yet discovered? Is the fact of its derivation and multiplication to be regarded as of no importance, and its mere matter, which after all *is* mov*ed* and chang*ed*, to be all in all? Might

not that very matter have been, and the very
forces manifested by it, have been for ages,
and might not similar conditions as regards
heat, light, moisture, and others, have per-
sisted, and for any length of time, and
yet no living stuff of any kind have been
evolved ?

But those who teach that life is force will
not begin by demonstrating the facts they rely
upon in connection with *simple* living things,
and gradually advance from these to the consideration of more complex beings. If we are
to learn anything about living things in general
surely our instructors should direct our attention
to the phenomena occurring in the lowest and
simplest beings, and during the earliest stage
of development of a complex creature, instead
of commencing upon a man, a full account of
whose life-phenomena cannot be comprised in
many volumes.

Now, why do so many philosophers exhibit
little inclination to begin by inquiring about the
simplest living things ? Why do our text-
books begin with the consideration of the complex physiology of the fully developed man,
instead of discussing, in the first place, the

simple physiology of simple living matter?
One would have thought that this was just the
point from which Mr. Mill, or Mr. Herbert
Spencer, would have desired to start. Here is
a thing increasing in size and then separating
so as to produce many like things. How does
it increase? Of course by drawing matter to
itself. But by virtue of what *property* does it
draw and select? What physical property
enables it to chose one thing and reject another?
How does it divide, and why does one portion
separate from another portion? Matter that is
alive first draws matter towards it, and then
this same matter separates itself, and at length
one portion moves away from the rest. Simple
phenomena, easily stated, easily demonstrated,
doubtless due to antecedent phenomena and
these to their antecedents; but is this the only
explanation that is possible? Doubtless if it
were not that physiology is embarrassed by
" natural difficulties " (Mill) these things would
have been explained by physicists long ago;
but if we argue as if we understood them
when we do not understand anything about
them, what do we gain by the process, however
successfully carried out?

A child can surely be taught that a little bit of soft transparent stuff takes up matter around it, which is not like it, and *converts* this into matter like itself, and so increases in size, and that it divides and subdivides, so that from one mass many masses result. This is what goes on in the development of the simplest living thing and in man himself. Not only is the process ccmmon to every known form of living matter, but it is peculiar to living matter, and is not known to occur in matter in any other state. And this matter came from pre-existing matter in a like state ! But the Harveian maxim, " Omne vivum ex ovo," says the Harveian Orator, " cannot perhaps be now maintained in its integrity ;" for science occupies itself with the " possibilities of occasional automatic generation" ! Men have indeed long been labouring at such possibilities of the imagination, but the experimental spontaneous ovum has yet to be brought forth.

Taking the marvellous range of living beings, from the simplest living speck which grows and multiplies under the most varying conditions, some having been regarded hitherto as incompatible with life in any form, to that living matter exercising in man the highest and most exalted function, which is destroyed if very slight change occurs in the complex conditions under which it has been ordained that it shall live—it seems extraordinary, considering the very confident assertions that have been made by those who advocate the physical doctrine of life, that not one single case to justify the assertion that vital actions are to be explained by physics and chemistry should have been brought forward.

It is idle and misleading to call heat, and light, and electricity, and motion "*vital forces*" when they are manifested in living beings, and *physical forces* when operating in the outside world. The *name* given to a force ought not to be changed according as the seat of its operation varies. But by this proceeding the advocates of the physical doctrine try to make people think that the *only* forces operating in living bodies are physical, and that these *are* the only forces which their opponents refer to when they speak of "vital power." The physical philosophers ignore entirely, or deny the phenomena which they cannot explain by physics, and seek to hide truly vital phenomena

in such phrases as " molecular changes," "physical relations," "·modified," " conditioned," and
many more.   In this way they succeed in en-
closing for the time, as it were in a thick mist,
the question at issue.

So, too, some speak of *physical* energy,
and of *chemical* energy, and of *vital* energy, as
if these were three forms of energy which had
been proved to be closely related to each other.
Of the chemical and physical forms of energy
something is known, but of the relationship of
the so-called *vital* energy nothing has been
proved.   We only know that the influence it
exerts is altogether different from that which has
been traced to physical and chemical energy.

Is it not incorrect to speak of the action
of a nerve, or muscle, as being due to *vital*
energy, seeing that the energy in question may
be simply physical and chemical, although it is
manifested in the tissues of a living being?
The energy is probably of the same *nature* as
the physical and chemical energy manifested by
non-living matter out of the body.   But if under
vital energy it is sought to include *formative
tissue-producing power*, an interpretation is given
to the term *energy* or force, which cannot be justi-

fied, as there is absolutely no reason for inferring that any mode of energy possesses formative constructive power. The idea of motion, or heat, or light, or electricity *forming*, or *building* up, or *constructing* any texture capable of fulfilling a definite purpose, seems absurd, and opposed to all that is known, and yet is the notion continually forced upon us that *vitality*, which does construct, is but a correlate of ordinary energy or motion. It is, however, obvious, that unless it can be shown,—1, that vitality can be converted into heat, or some other mode of force; or, 2, that some mode of force or energy can be made to assume the form of vitality, there is no sufficient reason for accepting the conclusion which has been held with such tenacity, and so unfairly forced upon public attention. The doctrine is indeed only a dogma resting upon assertion, and can only be entertained in opposition to every accepted principle of scientific observation and experiment. Education in the new physical philosophy has undoubtedly excited in the mind of many persons, certain prejudices from which they find it exceedingly difficult to emancipate themselves. The heavy penalties attached to the expressions of an opinion against physical force views, and the coercive persuasion employed in teaching molecular physics, tend to prevent free discussion, and enforce submission to a form of intolerance which is exhibited in the scientific writings of more than one member of the new school.

But surely it is very significant that every particle of living matter, of every sort known, should manifest phenomena of a particular kind —that is, should appropriate certain matters, and alter these, and grow and multiply by division, while no form of non-living matter has been discovered which exhibits any like phenomena.   It is said such matter will be obtained from the non-living *some day*, and that the view that non-living matter may do these things under "certain conditions" *not yet found out*, is in harmony with the "tendency of modern thought."   It is said that we are on the eve of discovering how to make living things from non-living matter—nay, that in a few instances this had actually been done.   But it is very remarkable that the drawings of the supposed new organisms said to be formed direct from the non-living, without parentage, so closely resemble certain organisms well known to us, and of very ancient lineage, that we may feel quite sure that the beings themselves, had they been compared, would not have been distinguished from one another. Indeed, the organisms supposed to have been prepared artificially, have in many cases been identified as a well known species, which had descended from its predecessors of the same kind in the ordinary vital manner. The supposed artificially produced organism will go through exactly the same phases of change as the one derived from a pre-existing creature.

It has been assumed that the actions of man and the highest animals differ in essential nature from those of the lowest creatures, but it would not be in accordance with the facts learnt by study, in any department of nature, to assume that the highest form of living matter is formed, or works, or acts upon principles totally different from those which obtain with respect to the lowest simplest kinds of living matter known. In the absence of positive evidence to the contrary, it would be dangerous in the extreme to assume that, for example, a monad may be built up anew from the non-living, while the man-germ, or the dog-germ, could not be so formed ; or that the white blood corpuscles, or the epithelial cells of man, grow and multiply and live like the lower, simpler organisms, while the cells of man's brain grow, and live, and act in some totally different manner. The proof of the working of a general law at one end of the scale of living beings, will soon be followed by a conviction of its application to phenomena occurring at the other. But the conclusion referred to would be directly at variance with the actual results of observation and experiment, opposed to the facts known in connection with development, and in the highest degree improbable. Confusion, not order, would in that case dominate in nature. Yet I will, nevertheless, freely admit that it is *possible* that the same living thing *might be* generated in two very different ways, but the facts at present known render it almost certain that no evidence will be obtained in favour of such an inference, and that as investigation advances the correctness of such a notion will be shown to be so very improbable that the idea will be abandoned.

The arguments which have been advanced in

favour of the genesis of living matter *de novo*, should be applicable to the formation of the marvellous sperm and germ elements, and the highest cells or elementary parts of man and the higher animals. If an organic particle can be formed in a solution of non-living matter, like a crystal, and assume the living state at the time, before, or afterwards, there is no good reason for concluding that the living particle from which the highest animals and man are developed, is produced upon different principles, or in obedience to different laws ; for neither in dimensions, nor in form, nor in composition, nor in any other essential character, property, or quality, to be demonstrated by physics, chemistry, or observation, does the one particle differ from the other.   Nay, no less an authority than Owen seems to accept this conclusion. He considers that all cells are really formed in this manner, and employs the term "formifaction" in speaking of the supposed deposition of cells and elementary parts from organic solutions.   Unfortunately, however, the "cells" supposed to be crystallised from a solution by "formifaction" existed long before the solution in which they are said to be formified was pro-.

duced.   The argument is faulty throughout.
The supposed facts are not facts, and the conclusion is necessarily fallacious.

The formation of a crystal in a solution is no
more analogous to the production of a monad
in a solution of organic matter, than the further
"growth" of the crystal is analogous to the
further "growth" of the monad, or than the
formation of a second crystal upon the first is
analogous to the development of a second
monad from that already existing.   The crys-
talline matter can be redissolved, and will
crystallise again as many times as we like, but
the monad matter cannot be redissolved and re-
formified, any more than a dog or a man can be
dissolved and then produced again from the
solution.   Neither man, nor any living thing,
nor any kind of living matter, can be dissolved,
for that which *lives* is incapable of solution.   It
may be *killed*, and then some of the products
resulting from its death may be dissolved, but
this is a very different thing *from dissolving the
living matter*.   Nor can the lifeless substances
which are dissolved ever be made to assume
again the form and character they once possessed.   Nor under any circumstances can the

living thing, once dead, be made to live again, even if no attempts whatever be made to effect its solution.

But let us mark the astonishing development which has recently occurred in regard to the new views. In the first place, Harvey's maxim, " Omne vivum ex ovo," has, we are told, become a mere " form of thought," and will soon be set aside in favour of such new and convincing conclusions as the following, which it is suggested may prepare our minds for the reception of the New Philosophy. Corresponding to the states, *living* monad, *dead* monad, have we not *living* crystal, *dead* crystal? For moving monad, growing monad—moving, growing crystal? For assimilating monad, assimilating crystal? For multiplying monad, have we not multiplying crystal? Does not the crystal, like the monad, proceed from an invisible germ? May not the crystal, like the monad, produce millions from a fluid holding in solution the materials of its pabulum? Does not the genesis of the monad, like the genesis of the crystal, depend upon a mere collocation of matter and force? Who, therefore, can refuse to believe that whenever the matter and forces are properly collocated,

living things will be formed *de novo?* And
since the earliest state of the matter of man
cannot be distinguished from the earliest state
of the matter of the lowest living forms, is it not
certain that when modern science shall have
discovered the complex collocations requisite
to produce such a result, man shall be able to
produce from a solution of non-living matter a
living germ,—if not of man, at any rate of one
of the higher animals? This, if placed under
the requisite conditions for development (which
are not understood at this moment, but will very
shortly be accurately defined) would, of course,
assume its characteristic form.   Indeed, if we
accept the statements that have been made, it
may be regarded as certain, that at no very
distant period an artificially evolved living being
will triumphantly proclaim its own independent
origin, and from its own experience relate to us
the story of its evolution, traced back through
various stages to its immediate construction out
of the non-living, when the lifeless molecules of
the inorganic so arranged themselves as to
develop from their lower forces the higher *vital
mode.*   Then will the sceptic regret his unbelief,
and joyfully become a convert to the New
Philosophy.

But now let us withdraw for a time from the dazzling speculations and fancies among which the physicist loves to dwell, and study for a while sober realities,—facts of observation,—in order to ascertain if from these we can learn anything that may assist us in forming an opinion as to the probable nature of the powers in living matter, the working results of which are so very different from those known of any forces acting upon matter in every other state. Let us carefully examine the actual structure of a tissue, and study carefully the manner in which its growth occurs, in the hope that, from the facts ascertained in the course of the enquiry, we may be able to decide if there is any good reason for believing that the physical doctrine of life rests upon a sure basis of fact, and determine if such a dogma as "the Sun forms the heart" is admissible.

It is a matter of indifference to me what tissue be selected for the investigation, and it is of no importance whether it be taken from an organism high or low in the scale of created beings,—whether it is animal or vegetable,— young or old. But as the *heart* has been adduced as an organ actually formed by the Sun,

D

let us take a small portion of its tissue and examine it. The thin muscular wall of the auricle of the heart of a young frog, or, still better, of the tree frog (Hyla), is well adapted for this purpose, as it is so very thin that no cutting or dissection is required in order to make it thin enough for examination even with the highest magnifying powers. It is so transparent that every texture entering into its structure can be seen in a properly prepared specimen.

The drawing in Plate I will give some idea of what can be made out. The network of branching fibres seen in every part of the specimen, and of different degrees of fineness, is muscular tissue. By the contraction of these the thin wall of the auricle is made to compress the blood contained in the auricular cavity, and force it onwards into the ventricle. But another network of still finer fibres, but not contractile, will be seen ramifying upon and amongst the muscular fibres. These are branches of nerves, through the instrumentality of which the contraction of the muscular tissue is effected.

At the lower part of the drawing, to the right, are seen some angular bodies of large size, occupying the spaces between the muscular

PLATE I.

THE MYSTERY OF LIFE.

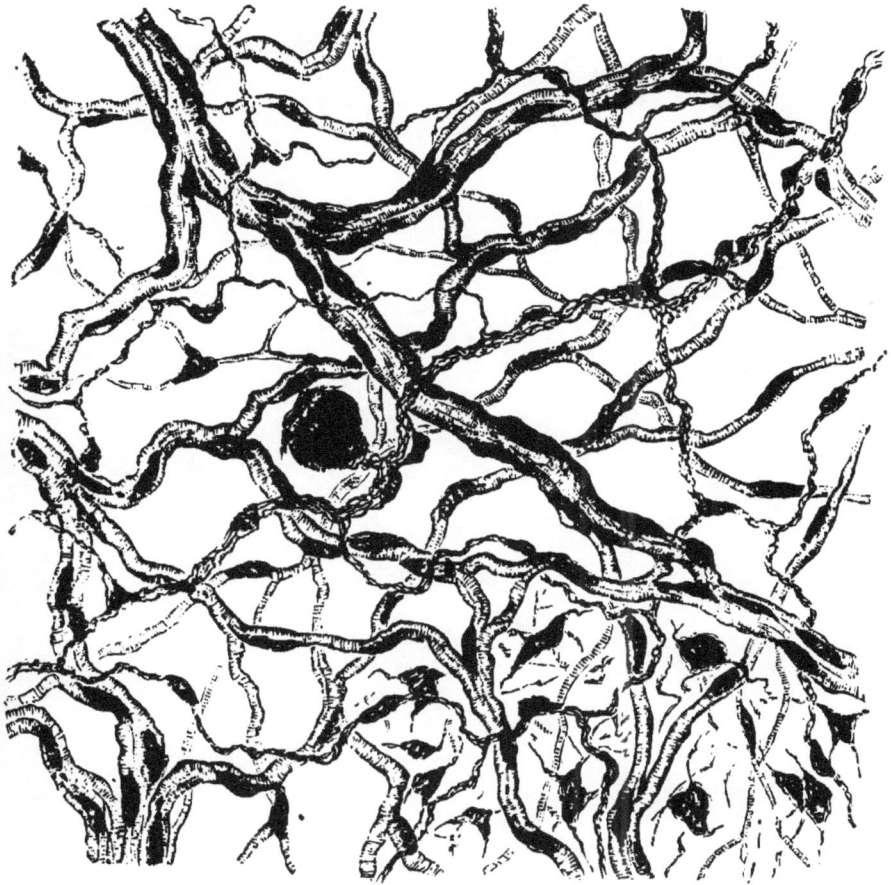

The thinnest portion of the auricle of the heart of the hyla, or green tree-frog, showing network of muscular fibres. The network of fine nerve fibres composed of still finer fibres is seen, ramifying amongst the muscular fibres. The connection of the finest nerve fibres with the ganglion cell, in the centre of the drawing, is well shown in the specimen. These fibres leave the ganglion cell, and as they pass to their distribution divide into bundles which pursue opposite directions. The connective tissue-corpuscles are represented in the lower part only of the drawing to the right. The bioplasm from which all these tissues have been formed is coloured red in the drawing. A portion of this figure is represented more highly magnified in Plate II, Fig 1, opposite p. 46.

$\frac{1}{1000}$ of an inch —— × 215.

[To face page 34.

fibres and the nerve fibres. These are the so-
called *connective tissue corpuscles.* They are
only represented over the small space indicated
in order that the arrangement of the muscular
and nerve fibres may be seen more distinctly.
These connective-tissue-corpuscles are concerned in the production of the transparent
passive fibrous or connective tissue which forms,
as it were, the basis texture of the auricular
wall—in which the more active nerve and muscular tissues are embedded. But instead of
discussing the characters of this passive texture,
we will consider briefly a few points of interest
in connection with the arrangement of the
active and more important tissues, the nerves
and muscles.

Connected with every tissue referred to, and
indeed every tissue in existence, are masses of
an oval, circular, or irregular form, consisting of
very soft structureless matter, which is coloured
red in the drawing. The masses were also red
in the specimen from which the drawing was
taken, in consequence of the tissue having been
exposed, immediately after death, to the action
of an ammoniacal solution of carmine. In this
way these bodies are obtained of a deep red

colour, while the *tissue* (muscle, nerve, &c.) remains perfectly colourless. By this remarkable action of an alkaline solution of carmine we are enabled to distinguish, and with certainty, every particle of living matter which *grows* and *forms;* from the passive lifeless matter in a tissue, which last possesses no formative power. The living matter in its normal state being perfectly colourless often escapes notice. Indeed, this most important constituent of living bodies has been passed over as of no importance at all. Its presence has been regarded in some cases as accidental, and every trace has been entirely omitted from many drawings of tissues in our text books. Consequently, the most erroneous views concerning the changes which occur during the formation of tissue and the nature and causes of them are entertained. I have therefore made new drawings of many of the tissues, for the purpose of showing this living matter, without which these tissues could not have been formed, and which is concerned in all the changes which the tissues and organs of the body undergo in health and disease.

The living matter differs entirely from matter in every other state, and I have called it *bioplasm, living plasm, germinal* or *growing matter.* Now in the drawing, masses of *bioplasm* or *bioplasts* will be seen embedded in, or in intimate relation with, the different tissues. In connection with the muscular tissue of the heart are very numerous elongated oval *bioplasts*, situated at short intervals from one another. More than five times as many of these exist in the cardiac muscular tissue as in ordinary muscular fibre. The activity of the changes of the muscular tissue of the heart is very great. It is always acting, and consequently new tissue is continually being formed to take the place of that which wears away. This is why the bioplasts are so abundant in the particular specimen of muscular tissue which I have selected for illustrating these remarks concerning the importance of bioplasm.

At an early period of development, the heart was represented by a collection of spherical bioplasts situated close to one another. These divided and sub-divided until a considerable mass had been formed. Some bioplasts were concerned in the formation of muscle, others were to produce nerve, and from others connective tissue only was to be developed. But at this early period all these bioplasts were spherical, all had been formed by division and sub-division from the same original bioplasm mass.   The nerve bioplasm cannot by any means known be distinguished from the bioplasm to take part in the formation of muscle, or either of them from connective tissue bioplasm.   Nor was there at this early period, when the basis tissue of the heart was being formed, a vestige of muscular or nervous tissue.   There was the living matter, by which alone the formation of such tissues is possible, but that was all.   As development advances, however, the little bioplasts move away from one another, and in their wake, or in the interval between them, *tissue* makes its appearance.   In this process the matter of the bioplast is changed, it loses its powers of growth and multiplication, and acquires the form and properties of tissue.   But only part of the living matter undergoes this change.   The bioplast does not disappear; it continues to take up nourishment and convert this into bioplasm, and at this, early period of life, faster even than its substance undergoes conversion into tissue. Thus the loss of bioplasm by transformation into tissue is more than compensated by the absorption of nourishment and the production of new bioplasm. These two processes, the formation of tissue by bioplasm, and the production of new bioplasm by the appropriation of nutritive matters, proceed at the same time.

What physical or chemical operation yet discovered can be compared with these nutritive and formative processes? Between the two sets of phenomena, *physical* and *vital*, not the faintest analogy can be shown to exist. The idea of a particle of muscular or nerve tissue being formed by a process akin to crystallisation, appears ridiculous to anyone who has studied the phenomena, or who is acquainted with the structure of these tissues. It is difficult to conceive how anyone who had thought over the facts, which are well known to every working student of physiology, could have succeeded in so misleading himself and others, as even to hint that the formation of tissue of any kind could be explained by physics and chemistry. There is not the shadow of argument founded upon fact, or upon the results of observation, to give countenance to such a doctrine. The idea that the ultimate molecules of living matter arrange themselves, or are arranged, by

virtue of the properties of the molecules themselves in such a manner that tissue results, is
supported by argument of about the same
degree of importance as that which might be
urged in favour of the theory that St. Paul's
and Westminster Abbey resulted from the properties of the stones of which those edifices
are built.

But further, it will be observed in the drawing
that the fibres of the muscular tissue interlace,
and that the fibres of the nerves interlace with
one another and with the muscular fibres in a
most intricate manner. It appears that there is
a certain proportion of nerve tissue to muscular
tissue, which is constant for each particular kind
of muscular tissue only. The nerve fibres form
an *uninterrupted network which exhibits no
endings*, which is everywhere distinct from the
muscular tissue, but the fibres of which cross
the muscular fibres at frequent intervals, and
are sometimes very close to them. Will any
one venture to affirm that the arrangement
which is delineated in the drawings, and which
can be demonstrated in many specimens, is to
be accounted for by physics and chemistry?
If so, let him make the attempt, let him adduce

reasons for the conclusion that the Sun, or force, or matter of any kind, could develop a particle of nerve or muscle, and show that there is some ground for the dictum, "the Sun *forms* the heart." But if this cannot be done it is time that those who with so little consideration have supported this outrageous assertion should withdraw it, and admit candidly that the heart is not formed by the Sun.

But, so far, we have only instituted a very superficial examination of the beautiful and delicate texture which constitutes the auricle of the frog's heart. We have not considered a tithe of the interesting facts presented for our contemplation by the study of this muscular and nervous tissue. Every nerve fibre is itself compound, and when we come to examine it under higher powers, with the aid of the advantages we now possess as regards methods of preparation, we discover that the finest fibres are resolved into still finer fibres, and it is doubtful if we can obtain a view of an ultimate nerve fibre, or if indeed it exists. Moreover, the fine nerves do not lie parallel to one another in the ramifications, but they interlace and coil spirally round one another; a fibre on

the right of a compound fibre crosses over to
the left, and then again traverses the branch to
reach the opposite side, and so on through the
whole length of the fine nerve fibres. This
arrangement is constant in all nerves, and is
observed in large trunks as well as in the finest
ramifications. Will anyone believe that a fact
so constant is meaningless, a result of accident
to be explained by subtle influences, or natural
selection ? Shall we account for the fact by
attributing it to the operation of a physico-spiral
tendency, or dismiss it with the announcement
that the law of nerve reticulation will be discovered ere long ? But it is marvellous how
difficult it is even to obtain assent to an important but simple fact of observation of this
kind. Every student who has dissected a
nerve knows that the nerve fibrils cross and
intertwine in the trunk. A low magnifying
power only is required to convince him that the
same fact is observed in the smaller trunks.
Careful examination of properly prepared specimens will prove that the disposition is the same
in still more minute ramifications ; and I possess
numerous preparations in which the same point
may be demonstrated in the most minute nerve

ramifications we can see under the highest powers yet made (the $\frac{1}{25}$ and $\frac{1}{50}$ magnifying respectively 1,800 and 2,800 diameters). The arrangement in question is represented in a great number of my drawings published during the last thirteen years. And yet some of the most careful observers seem unacquainted with the fact, and not aware that attention has been directed to it. In truth they contradict observations, the correctness of which can be put to the test by the examination of a good specimen in five minutes.*

* Some elaborate and otherwise very correct drawings in the very last number of Max Schultze's "Archiv," are defective in this particular. In the very large drawings the nerve fibres composing the fine compound ramifications are represented by *parallel lines* ("Die Flughaut der Fledermäuse, namentlich die Endigung ihrer Nerven," von Dr. Jos. Schöbl, in Prag. Siebenter Band, Erstes. Heft, 1870). This paper is a very valuable one, and is particularly interesting to me, inasmuch as the author confirms many of my own observations concerning the ultimate distribution of nerve fibres, which were made several years ago, and were at that time discredited in Germany. The networks of fine fibres described and figured in several of my papers, and illustrated in Plates I and II of this essay, are represented by Schöbl in three colours over a very extensive surface of tissue. The author's delineation of the distribution of nerves to muscle also accords with my own. Dr. Schöbl will,

How is this wonderful interlacement and intertwining of fine nerve fibres, of larger nerve fibres, and of the largest compound trunks, in every part of the periphery, and in every central nerve organ, to be accounted for by any physical hypothesis ? No such arrangement can be brought about artificially, except by the interweaving of threads which have been formed or spun in the first instance, and we know that this is not the manner in which the nerve trunks and plexuses are produced, for such a process is rendered impossible by the conditions under which the formation of nerve fibres proceeds in living beings. By admitting *vital power*, it is, however, possible to account for the phenomenon without ignoring any of the facts which can be demonstrated during the course of development ; but the question is too extensive to consider in this place.

We have, however, as yet only subjected our

I am sure, excuse me for making these remarks upon his recent work. By criticising each other's observations in such particulars, we shall promote that exactness and care in recording observations, and making drawings, which so surely promotes the true advancement of that department of science in which we labour.

specimen to a very incomplete examination, and have found that physics will not fully account for any of the facts revealed by observation. I might further describe the wonderful structure of the beautiful ganglion cells represented in Plate I and Plate II, figs. 2 and 3, by which the rythmic contraction of the muscular fibres is effected, and with which the nerve fibres are continuous, but by so doing I should probably tire the reader with too many minute details. The conclusion, however, already deduced would again be arrived at,— that neither the structure, nor the arrangement, nor the position, nor connections of these little nerve organs could be accounted for by physics, nor their composition or action explained by chemistry. Not even the connective-tissue-corpuscles have been formed by force, nor do they grow or act by physics and chemistry.

Every particle of tissue, represented in the drawings, is the result of changes which have occurred in previously existing living matter. The evidence of the origin of the tissue is as distinct and certain as that which leads us to conclude that the formation of the lifeless shell cast upon the seashore, is due to changes effected during the life of a living animal.

The *action* of such tissues as nerve and muscle is determined by their structure and composition, which are a direct consequence of the influence of vital power upon the particles, of matter of which they consist. We cannot, therefore, show even that the *action* of these tissues is due to physical and chemical changes only, for not only is the action dependent upon the *structure*, but the maintenance of the structure in a state fit for action is effected by the living matter, or bioplasm, which exists in greater or less proportion, and is more or less active during every moment throughout life.

But if any advocate of the physical doctrine thinks I have unfairly selected a highly complex structure for the express purpose of bringing out too strongly the objections to his favourite hypothesis, let him select some simpler tissue, and explain if he can its formation according to his own views. Nay, if it be preferred, let the phenomena of the simplest living organism in existence be taken as the subject for discussion. It seems to me clear that facts are more favourable to the *vital* than to the

PLATE 11.

THE MYSTERY OF LIFE.

Fig. 1.

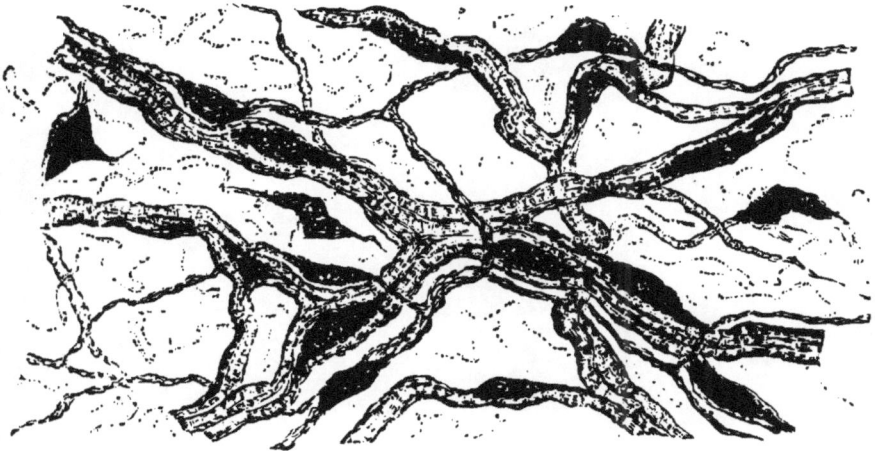

A portion of the same specimen as that figured in plate I, magnified 700 diameters, showing the relation of the masses of bioplasm to the nerve and muscular fibres, which have been formed by them. One or two connective tissue corpuscles are also represented.

Fig. 2.

Fig. 3.

Ganglion cells. Showing straight and spiral nerve fibres, and their course, in *opposite directions*, in the trunk of the nerve. The bioplasm of the ganglion cells and nerve fibres is also shown. From the hyla or green-tree frog. × 700. p. 45.

Ganglion cells and nerve fibres with the bioplasm which has taken part in their formation. An arrangement similar to that represented in this drawing and in Fig. 2 is seen upon all the nerves distributed to the heart, lungs, liver, kidneys, and intestines. From the newt. × 130. p. 45

$\frac{1}{1000}$ of an inch ———————— × 700.

$\frac{1}{1000}$,, ——— × 130.

[To face page 46.

physical hypothesis of life. Neither physicist nor chemist can explain by physics and chemistry the increase in size, or the division and sub-division of the tiniest monad, or the lowest microscopic fungus. It is time that his incapacity to do so were distinctly admitted and definitely acknowledged, and that persons unlearned in the details of life science should no longer be encouraged in the belief that living things are force-constructed mechanisms, and that life has been evolved from ordinary matter in which there was no life.

Next, let me inquire if the facts so familiar to us who are daily brought face to face with the various remarkable changes occurring in the tissues of man's body in disease are to be explained by physics and chemistry. It would surely be difficult to find remarks having any pretension to scientific accuracy more pitiful than many of those which have been advanced as physical and chemical explanations of the phenomena of disease. To fully explain any disease whatever by mechanics must be impossible, unless the phenomena of health are also susceptible of mechanical explanation. What can be more vague and inconclusive than the chemical theories which have been offered as " explanations" of the phenomena of fever and inflammation? In cases in which the patient is dying of suffocation, when one lung has ceased altogether to breathe and the other is almost obstructed, we are told that, nevertheless, the fever depends upon *increased oxidation*. If the temperature rises, as it does in some cases many degrees, *after death has occurred*, and continues to rise for some hours after heart and lungs have ceased to act, we are expected to assent to the dictum that the fact is due to increased oxidation. But it is obvious that rise in temperature in such cases is associated with diminished, instead of increased, access of oxygen, and the chemical theory of animal heat can only be accepted if the most important of the well established facts be ignored. In the same way mechanics and chemistry utterly fail to explain the phenomena of an ordinary cold, or a common headache. The effects of a sting, or those following the bite of a gnat, or a flea, no more can be accounted for by physics or chemistry than those resulting from the poison of the rattlesnake or cobra, or from the introduction into the system of a few drops of hydrocyanic acid, a little strychnia, or the fraction of a grain of nicotina.

Then there are the marvellous but familiar facts in connection with contagious, self-propagating diseases.* Is the poisoned dissecting wound a mere chemical or mechanico-chemical phenomenon ? To what chemical actions does it exhibit the slightest analogy ? and who feels satisfied when he is told that the speck of vaccine lymph is but a bit of albumen in a state of rapid chemical change, which induces chemical change in the fluids and tissues of the body ? This sort of phrase is often advanced in place of explanation, but not one of the terms of the proposition has any intelligible meaning attached to it. The whole question is begged, and it is expected that the inquirer will be silenced by the learned words which, in his innocence, he may fancy must have some deep or mysterious meaning that his poor intelligence does not comprehend, or his too scanty knowledge enable him to interpret. Can the changes in one of the little mucus corpuscles from the mucus of the throat be explained by physics, or the movements of the white blood

* See " Disease Germs, their Real Nature," 1871.

E

corpuscle, or the growth of pus by chemistry?
Does the poison of measles differ from that of
small-pox chemically or mechanically, or in
some other way? Is the escape of one indi-
vidual and the invasion of another to be accounted for upon mechanical principles, or has
the poison a chemical affinity for one, and an
instinctive antipathy to another?

In a common cold there is increase of *living
matter*, but, like the fact, the substance has
been ignored by physicists, chemists, and the
chemico-mechanical school of medicine. The
increase of living matter occurs in all fevers
and inflammations, but by the chemist this fact is
neither admitted nor recognized. The doctrine
of correlation will not assist us in interpreting
the phenomena. There is not a disease in
which it cannot be shown that *vital*, as distinct
from *physical* and *chemical*, changes are at work.
Every remedial measure we employ does good
or harm in bringing about conditions which are
favourable or unfavourable to the growth and
multiplication of such soft, transparent, semi-
fluid, living, moving matter, quite irrespective
of any merely mechanical or chemical influence
it may exert.

Physiology and medicine are not branches of physics, and, like many other departments of human knowledge, cannot be comprised in mechanical philosophy. If the facts of health and disease could have been explained by chemistry and mechanics we should have had the explanation long ago. It is now quite time that chemists and physicists admitted their inability to account for them, and ceased to exhibit hostility towards those who refer the phenomena peculiar to living beings to the influence of vital power until some sufficient physical explanation shall have been given. At present no one who uses his reason rightly can admit that the facts of the case justify him in looking upon his organism, or any other organism, as a mere force system, the normal equilibrium of which may be somehow mechanically or chemically disturbed in disease— as a clock crystallised from its mother liquor, having the property of ticking monotonously for a time, capable of receiving a new wheel or spring if either be broken or lost, but when damaged by dirt or rust, or worn out by age, good for nought but to be cast into the melting-pot, where its " properties " become " modified," its structure destroyed, and its individuality discharged for ever.

After having been educated to ignore half the facts of existence and abandon all the hopes, it is possible people might be persuaded to believe that all the phenomena of life and of man are fully accounted for by physics and the doctrine of the conversion of force. But by a special course of instruction only could the mind be prepared for the acceptance of the view that the construction of the most elaborate organism, the marvellous acts performed by it, the capricious vacillations of the lowest human will, as well as the grandest creations of the most perfect human intellect, are but indications of the quivering oscillations of a force wave, the varying intensity of which is determined by the ever-changing conditions occasioned by its own eternal undulation. But inaccurate generalization and vague assertion have been carried still further. It is gravely contended by some that our part of the universe is undergoing degradation which must progress, and that, in consequence, life on our globe will ere long cease. The fire of the Sun, the great *preparer of our food (!)*, as well as the *builder of our tissues*, is, we are told, gradually going out, and, unless more fuel is supplied to compensate for the excess of expenditure over income, there is no hope for life. But after having dilated upon the colossal physical changes which certainly will lead to the extinction of all life, our teachers next admit that we have yet much to learn concerning the data upon which is grounded this definitely stated opinion that life will cease, &c. ; in short, that the portentous conclusions they have themselves deliberately deduced are, after all, but mere conjectures, which are certainly not supported by any facts of science yet discovered. Nor can the physicist adduce reasons for supposing that it is at all likely that within the next century the facts required to justify the expression of these and other recently evolved fancies will be at his command.

Dr. Gull, like many who disapprove of the *vital theory*, admits that he cannot fully explain vital phenomena. Vitality is, then, after all, a *mystery*. But some of us are convinced of the truth of facts which justify us in concluding that the mystery is to be accounted for only by supposing an agency, force, or power of an order different from that in which the forces of

the non-living world are included; while others
maintain that life will eventually prove to be
but another mode of the ordinary forces of
matter. For my part, I am ready to abandon
altogether the idea of vitality, and to dismiss it
with other ideas, considered by the new school as
mere prejudices imbibed during the irresponsible
state of childhood, as soon as convincing evidence of error shall be adduced; but I refuse
to give up these for the threats or gibes of a
school whose tenets rest upon the mere authority of modern assertion, and whose forcible
dicta, however determined and arrogant, are
justified neither by reason, nor by observation,
nor by experiment.

There is a mystery in life. A mystery which
has never been fathomed, and which appears
greater the more deeply the phenomena of life
are studied and contemplated. In living centres,
far more central than the centre as seen by the
highest magnifying powers—in centres of living
matter where the eye cannot penetrate, but
towards which the understanding may tend,—
proceed changes of the nature of which the
most advanced physicists and chemists fail to
afford us the faintest conception. Nor is there

the slightest reason to think that the nature of these changes will ever be ascertained by physical investigation, inasmuch as they are certainly of an order or nature totally distinct from that to which any other phenomena known to us can be relegated.

Lastly, it may be well to consider if our own will, feelings, thoughts, emotions, hopes, desires, can be expressed in force terms, or measured by force standards. We are told that the nervous tissue is highly *plastic*, the plasticity being no doubt due to the property of the " clay " of which it is made, by virtue of which " it is not only capable of receiving and registering the impressions made upon it, *but of acquiring an instinct for complicated acts,*" and *this*, Dr. Gull tells us, is " the physical basis of education and of even morals !" Now where, I would ask, is the lifeless clay, the inanimate plastic substance, which acquires an instinct ? Does not this very " plasticity " of the nervous system, so different from the " plasticity " of inorganic substances, remove it at once from the category of the non-living ? But nerve-plasticity may be yet another undiscovered correlate of clay-plasticity, and both of them

but converted primary energy; in which case
morals may be regarded as the outcome of a
highly plastic physical basis.

The organs of the senses receive physical
impressions. But how does this fact give any
support to the conclusion that these organs are
themselves the result of mere physical and
chemical changes? The ear or the eye *formed*
by physics, *because* one distinguishes the vi-
brations of sound and the other those of light!
Now, that such views should be entertained at
all is but evidence that he who holds them is
not acquainted with the structure of these
wonderful organs. It is most unreasonable
on the part of any one to allow such an opinion
to pass current so long as the steps by which
the arrangement of the simplest nerve plexus,
which we can demonstrate easily enough
(Plate I), was brought about, continue unknown,
nay, while the actual mode of arrangement,
and termination of the nerves in the simplest
terminal nervous organ is admitted to be doubtful. But while there is so much yet to be
discovered as regards the mere structure of the
simplest nervous mechanism, what must we say
of those who profess to be able to tell us the

precise nature of the actions of the highest parts ? Without being able to give us an idea of the structure and arrangement of the *apparatus*, they do not ·hesitate to assure us that its *action* is *mechanical and chemical*, and that the marvellous thinking instrument, whose intricacies have never been unravelled, is merely plastic matter, formified from its solution after the manner of the deposition of a crystal from its mother-liquor.

Man, as well as man's brain, we have been told, is formed by " evolution." His organs result from " evolution," and the higher mental faculties with which he is endowed, like the instrument of which these are the supposed function, are " evolved " from the more simple. So that a complex structure may be " evolved " from a simpler structure, and a complex action from a more simple action.

But " evolution," like many other terms employed in the science of our day for the purpose of accounting for phenomena, has had no definite meaning assigned to it. To say that a thing has been formed by " evolution " conveys information less definite and less correct than is conveyed by the statement that it has been derived from a pre-existing living thing. The formation of tissue has been attributed to "vacuolation" and "differentiation," and these polysyllables have lately been superseded by the still more vague terms "subtle influences," and "external conditions," and "sundry circumstances." And it has been affirmed that to "the primitive properties of the molecules" and "natural selection" may be referred all the varying forms and structures known to us as well as all the phenomena of the living world. But such terms explain nothing. By their use further enquiry is discouraged, and the mind bent upon investigating the secrets of nature is misled at the very outset. Can any one of these very pretentious phrases be resolved into anything more than the statement of a fact or facts in the form and language of an explanation ? Natural selection is the formation of species, and species are produced by natural selection. Crystallisation is the formation of crystals, and crystals are produced by the operation of crystallisation. Tissues are formed by differentiation, and differentiation is the formation of tissues ; and so on. But whether formation be attributed to "subtle influences" and

"sundry circumstances," or to evil influences,
witchcraft, or the influence of fairies, can
surely be of very little consequence.   By such
explanations, especially if conveyed very em-
phatically, and with authority, the unlearned
may be astonished, and pleased, and confused,
and imposed upon, but those who put forward
such explanations do not convey information,
and instead of promoting the advance of natural
knowledge they retard real progress.

Dr. Gull, with many more, at present shrinks
from regarding mind as correlated force, and
therefore does not at this time look upon man
as a mere mechanism.   But unless it shall be
shown exactly where the lower forms of life
are marked off from the higher, this is a position obviously untenable.   The man-germ has
no more mind than the dog-germ or the cab-
bage-germ.   At what period of development,
then, according to the view above referred to,
does the man-germ become distinct from all
other beings, and acquire those properties which
make man "a being apart?"   At what period
of his being is that "immeasurable and impassable gulf" excavated, which is supposed to
separate him so decidedly from the rest of

creation, and by what method of investigation is the gulf to be rendered evident to the senses ?

On the other hand, mind itself is, by many who understand the force of logical reasoning, considered to be but the result of molecular changes in nervous matter, and the arrangement of this nervous matter is supposed to result from the operation of certain *complex conditions*. Chemical action, it is held, may be convertible into mind, just as heat is convertible into electricity or motion. Only the conditions required to bring about the first conversion are much more complex than those by which the other may be effected. *Memory*, it may be said,—nay, it has been said,—is but the capacity to register the effects of impressions, which nervous matter enjoys in common with every organic element, and certain inorganic matters, including *stones and crystals!* Nor is there a faculty of the mind which cannot be disposed of in the same way. Force *forms* the brain, which *converts* force into mind.

But in all these notions the act of formation, the cause of formation, and action after formation is complete, are confused together. It is held that the organ which changes force has itself been constructed by force. Force is conditioned by the apparatus it has built up. Force is the architect, the director, the builder, and force is afterwards directed, changed, and modified by the working of the machinery it has designed, constructed, and made. Force is that which conditions, and that which is conditioned. Force forms the instrument which correlates and is correlated by it. It is at one time that which produces the correlating apparatus, and at another is itself correlated by the results of its own constructive power. The constructor is a correlative of the work performed by the mechanism he has produced. The *artificer*, the *machine*, and *the work* done by the machine, are then all correlative!

But does not "life" exist before brain and nerves, the instruments of mind, are formed? If then "life," which manifests itself in man's organism before mind is evolved, which sometimes exists independently of mind, but without which mind could not exist, be a correlate of heat, it must be a correlate very different from mind. The difference, it will be said, is due to the difference of the molecular machinery which effects the conditioning of the forces. But the machinery has been formed by force, so that we must assume that there is not one new correlative of motion, life, to be discovered and produced anew in the laboratory ; but others, of which mind takes the precedence. For the transformation of primary energy into mind, ·life is necessary, for life invariably precedes mind. Life may be manifested without mind, but the manifestation of mind without life is impossible, even in the conception of the physicist.

It is in the present state of knowledge simply astounding, that reasonable people should accept the dogma that life is a correlative of heat. There is not more than the shadow of foundation for such a view. But that mind should also be received as another correlate, only proves that few persons think about the mental actions going on in their own organisms, and that dogmatic assertion of one kind is as powerful to influence them as was that of another kind to mislead their forefathers.

Those who do not go quite so far, but argue · concerning the *possibility* of life being a correlate of ordinary force, should bear in mind that nothing can result from the mere assertion that *vital* force *may be* another form or mode of heat or motion, unless facts and arguments can be advanced in support of the supposed possibility. The assertion has been repeated hundreds of times, but the arguments which have been adduced hitherto, have been shown to rest upon no secure foundation. At the same time I I would say, " By all means let the idea of *vital power* be upset, for once and for all, if this can be done." I hold it because I cannot escape from it, because the facts I know, cannot be explained without the hypothesis. .

The most sanguine physicists are *perfectly sure* that *thought* and *life* itself will *some day* (!) be summarily transformed into a new undiscovered correlate by the might of unthinking force. This is to happen as soon as the proper structure of the conditioning machinery, which is to effect the change, shall have been determined, and this is to be effected by force which is at the same time condition, conditioned, and correlate. But so far the transformation of life into force, or force into life, has not been effected. Nor has any one yet succeeded in showing that the fulfilment of this *possibility* is near at hand, or that it receives support from any newly discovered facts, or recently conducted observations or experiments. I am quite ready to be *taught*, but I cannot submit to be *forced* into confusion by force, while I retain vital power to resist. During the last twelve years numerous facts, elucidated in the course of careful microscopical investigations on the tissues of plants and animals, which have not been called in question, tend to establish upon a firm basis the doctrine of "vitality;" or at the least indicate that the phenomena peculiar to living beings are due to the working of some special power capable of guiding, and directing, and arranging ordinary matter, but in no way emanating from, or correlated with, the ordinary material forces. I cannot but conclude from my investigations that the living is separated from the non-living by an impassable barrier—by a gulf that will not soon be bridged over; that matter and its ordinary forces and properties belong to one category or order; and that creative power, and will, design, and mind, and life, ought to be included in a very different order indeed.

In conclusion, I submit that the arguments advanced by Dr. Gull, and others, do not show that the opinion that life "is a power entirely different from, and in no way correlated with, matter and its ordinary forces," is untenable. Neither can it be held that the reasoning advanced by him in any way justifies the acceptance of the hypothesis that life is correlated force. This physical doctrine restricts advance and retards scientific progress. It cannot be accepted without straining, to a degree quite unwarrantable, arguments which are based upon conjectures instead of facts, and denying or ignoring many well authenticated facts which have resulted from numerous observations, and can be confirmed without difficulty. On the other hand, the theory of "vitality" helps us to explain many phenomena otherwise inexplicable at this time, while it is not incompatible with any of the truths of physical science.

The Physical Nature of Vital Energy.

The following observations were published in the " British Medical Journal," October 29th, 1870, in answer to some remarks made by Dr. Ferrier in favour of the physical doctrine of life and against my views on vitality:—

The conviction that it is "*only* by recognising the *physical nature* of *vital* energy that we can ever hope to establish therapeutics on a firm and sound basis," has perhaps led Dr. Ferrier*
to express himself rather decidedly against some views which I ventured to put forward some years ago, but which I am ready to give up as soon as convincing evidence shall be adduced in favour of the physical doctrine of life.

If Dr. Ferrier will explain what is meant by " molecular organization " and " molecular machinery," he will serve the cause he has at heart far better than by attacking me ; for, as he must have gathered from many of my remarks, I am quite as anxious for light as any one can be. What I desire is to learn in what

* Introductory Lecture on Life, &c. ("British Medical Journal," October 22nd.)

particulars the "*living*" resembles and differs from the "*non-living*." I am quite ready to admit that one living thing is only some other living thing, or dead thing, or non-living thing, "variously modified" "under sundry circumstances," by " subtle influence ;" but I should certainly like to have the meaning of these very ambiguous phrases explained. A man may be said to be only dust "variously modified ;" but consider what is comprised in the " variously modified ! " And perhaps I may be permitted to ask why, if it is right to attribute the marvellous phenomena of nutrition to "subtle influences," am I to be condemned because I prefer to employ provisionally the simple term " life," or " vitality," or " vital power" ?

It is *possible* the " molecular machinery " which is supposed to condition the physical force and convert it into the *vital* mode may be discovered ; but at present it is absolutely unknown. It has never been seen, and no one has yet told us what it looks like even in his imagination.

But I will admit that it is *possible* such machinery *may be* beyond the microscopic limit. The imagination of *highly gifted persons may* be able to conceive the structure and mode of action of the molecular machinery of the existence of which they are perfectly certain, although it has not yet been rendered evident even to their sense. Nay, I will admit further, that a *sufficient* intelligence might be able to predict, from ·the properties of its component parts, the character which the offspring of any given piece of "molecular machinery" will assume after it has continued to grow and multiply, say for a thousand years. But do such suggestions enable us to unravel the mystery of the life of even the simplest thing now alive, or to determine in what particulars a *living* particle differs from the same particle *dead ;* or why a portion of a mass of living matter moves upwards as well as downwards, or in what manner it takes up non-living matter, and communicates to this its own properties, and divides into separate portions, every one of which possesses equal powers ? It may be answered,— " These phenomena are due to the properties of the molecular machinery which has long been known to exist in the imaginations of highly gifted persons ; and, although as yet no one has succeeded in actually producing such

machinery artificially, the efforts of the philo-
sophic imagination tend towards such a con-
summation!" But surely no observer, no
worker at science, will feel satisfied with such
statements as these; and a few will probably
agree with me in thinking that, although it be
in a sense *un*philosophical, it is neither incon-
sistent nor absurd, to entertain the opinion that
the vital phenomena of living matter, which was
derived from pre-existing living matter, are due
to a peculiar power. At the same time I object
to accept the view that the action of a steam-
engine, which was not produced by a pre-existing steam-engine, is due to a "steam-engine
principle;" and I confess it appears to me very
extraordinary that many advocates of the
physical theory of life cannot be convinced that
the analogy they draw between a machine—
which does not make itself, or grow, or multiply
—and living matter, which seems to do all these
things, is so very slight as to be beyond every
limit except that of the fancy. If those who
support the view which Dr. Ferrier so strongly
advocates could explain by physics and chemistry (a) *the movements*, (b) *the growth*, and (c)
the division of any particle of living matter of

any organism in this world, they might have
some excuse for the very positive statements
they make about the *physical theory of life*.
But they have not explained these things, and
they know they are not to be explained by
physics.

People are beginning to doubt whether, after
all, living things are really so like machines and
crystals and physical bases, and complex albu-
minoid matters in a state of rapid chemical
change, as they have been led by the disciples
of the new philosophy to believe them to be.
And people are also beginning to doubt if those
who have spoken so positively on the physical
side really know much more than any one else
knows about the nature of life; although, from
their very decided manner, it was natural to
believe they possessed very peculiar and perfect
knowledge of the secret.

Whether the physical theory of life would
have resisted much better the "*furious on-
slaughts*" that have been made against it, if
some other course had been pursued, is a matter
of opinion; but it is quite certain that some of
the strongest supporters of the doctrine are
modifying their views, and are preparing to

modify them still further. Those who have watched for ten minutes, under a high magnifying power, the varied movements of living matter, and have thought a little over the question of the nutrition of that living matter, will not easily be brought to believe that such phenomena are due to physical and chemical changes only. The number of such observers increases daily.

HARRISON and SONS, Printers in Ordinary to Her Majesty, St. Martin's Lane.

WORKS BY THE SAME AUTHOR.

8th Thousand, 21*s.*

HOW TO WORK WITH THE MICROSCOPE.

The Fourth Edition, very much enlarged.

This work is a complete manual of microscopical manipulation, and contains a full description of many new processes of investigation, with directions for examining objects under the highest powers.

With Seventy Plates, including many new Figures, some of which are coloured.

CONTENTS :—

"The Author, both in the text and in the explanations of the engravings, has endeavoured to restrict himself, as far as possible, to giving hints and directions which may be practically useful to the student while he is at work." —*Extract from the Preface.*

HARRISON, PALL MALL.

THE PHYSIOLOGICAL ANATOMY & PHYSIOLOGY OF MAN.

By ROBERT B. TODD, WILLIAM BOWMAN, & LIONEL S. BEALE,
Fellows of the Royal Society.

Being a New Edition, by Dr. BEALE, of Vol. I. of the original Work.
Part I., with Plates, now ready, 7s. 6d.   Part II., *nearly ready.*
LONGMANS & Co.

1s., Free by Post.
ON MEDICAL PROGRESS. IN MEMORIAM: ROBERT BENTLEY TODD.

Cloth, 8vo., 16s.
CLINICAL LECTURES,
By the late ROBERT B. TODD, M.D., F.R.S.
Second Edition.   Edited by Dr. BEALE.

Fourth Thousand, 16s.
THE USE OF THE MICROSCOPE IN MEDICINE.
FOR PRACTITIONERS AND STUDENTS.

Third Edition, 25s.
KIDNEY DISEASES, URINARY DEPOSITS, AND CALCULOUS DISORDERS;
Including the Symptoms, Diagnosis, and Treatment of Urinary Diseases.
With full Directions for the Chemical and Microscopical Analysis of the Urine
in Health and Disease.

The Plates separately, 415 figures, 12s.   The Text, pp. 500, 15s.

Recently Published, 46 Figures, 5s.
I.—ON THE STRUCTURE AND FORMATION OF CERTAIN NERVOUS CENTRES,
Tending to prove that the Cells and Fibres of every Nervous Apparatus
form an uninterrupted Circuit.

II.—INDICATIONS OF THE PATHS TAKEN BY THE NERVE CURRENTS
As they traverse the Caudate Nerve Cells of the Cord and Encephalon.
One Plate and 4 Figures, 1s. 6d.

ON DISEASES of the LIVER and their TREATMENT.
A Second Edition, much enlarged, of the Author's Work on the Anatomy
of the Liver.   Numerous Plates.   [*The First Part shortly.*

3s. each Number.   Subscription for 4 Numbers, 10s.
THE ARCHIVES OF MEDICINE.
A Record of Practical Observations and Anatomical and Chemical Researches
connected with the Observation and Treatment of Disease.
Edited by Dr. BEALE.
Vols. I. and II., 15s. each; III., 11s.; and IV., 13s.
The first number of Vol. V. now ready.
All Communications to be addressed to the Editor.

J. & A. CHURCHILL.

NEW ILLUSTRATED WORKS.

ON VITAL PHENOMENA, AND ON THE MINUTE STRUCTURE OF LIVING BEINGS IN HEALTH AND DISEASE.

These Works are all illustrated with Coloured Plates, and treat of subjects of great importance to the progress of Physiology and Medicine, technical terms being as much as possible avoided.

OF THIS SERIES THE FOLLOWING ARE ALREADY PUBLISHED.

Second Edition, very much enlarged, 6s. 6d.

1. LIFE, MATTER, AND MIND; OR PROTOPLASM.

With Original Observations on Minute Structure and numerous New Coloured Drawings.

** This work is partly original and partly controversial.

Four Coloured Plates, 3s. 6d.

2. DISEASE GERMS: THEIR SUPPOSED NATURE.

An Original Investigation with the aid of the Highest Powers yet made.

[*Now ready.*

Twenty-four Plates, 16 coloured, 8s. 6d.

3. DISEASE GERMS: THEIR REAL NATURE.

An Original Investigation on the Minute Pathology of Contagious Diseases.

Two coloured Plates, 3s. 6d.

4. THE MYSTERY OF LIFE.

An Essay in reply to DR. GULL's Attack on the Theory of Vitality in his Harveian Oration for 1870.

Nearly ready, with 6 Coloured Plates.

5. LIFE-THEORIES: THEIR INFLUENCE UPON RELIGIOUS THOUGHT.

PREPARING.

DISEASE: ITS NATURE AND TREATMENT.

ON BIOPLASM, OR THE LIVING MATTER OF THE TISSUES AND FLUIDS OF LIVING BEINGS.

With Numerous Coloured Plates. An Introduction to the Study of Physiology and Medicine. Being the first Course of Lectures delivered at Oxford by direction of the Radcliffe Trustees.

ON INFLAMMATION: ITS NATURE AND THE PRINCIPLES WHICH SHOULD GUIDE US IN TREATMENT.

** *All these Works contain the results of the Author's original investigations. They are illustrated with upwards of 2,000 Engravings, copied from the actual objects, all of which have been drawn on wood by the Author himself or under his immediate superintendence.*

London : J. & A. CHURCHILL.

www.ingramcontent.com/pod-product-compliance
Lightning Source LLC
Chambersburg PA
CBHW021955190326
41519CB00009B/1272